Mini Corpus Venn 3

Contact Information:

diogodesouza7@gmail.com
diogodesouza7@hotmail.com

Life is like music. It has a rhythm, and it is like a flow of multiple events and experiences that leads us through the dreams of our everyday existence. This flux of thoughts is what causes us to gain experience and discover through the patterns of causes and effects the meaning and the workings of the cosmic order that sustains the universe which is our home.

Life is Music

The greatest gift is to exist and to do what is always right.

Having fun while living what is right and rejecting what is wrong is the way of being in perfect harmony with the purpose of what is the true meaning of joy.

To treat everyone with carefulness and respect while maintaining also oneself with a high value leads to a healthy relationship.

It is important to follow a strict and noble law by performing noble actions and like a mirror, your life will pay you back with peace, joy, and more wisdom. If you do what is not fair, and hide your evil while still acting evil, the mirror will condemn you which is the way to failure.

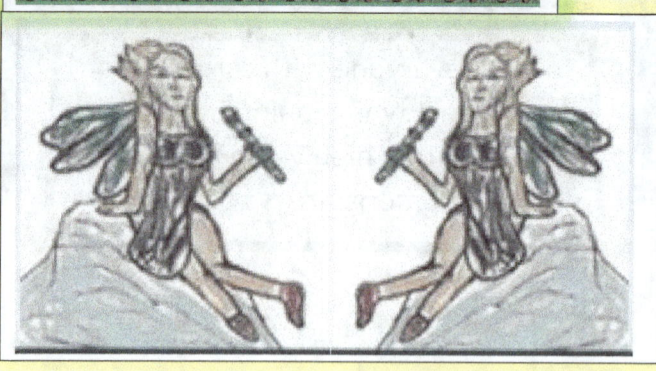

Your profit mirrors your actions. Payday happens everyday.

It is important to work hard but equally important is to take a long break and rest so that the hard work may continue bearing great fruits.

We can only solve complicated problems if we make simple and small steps, while keeping a look at the situation in trying to see the problem as being simple. It is by simplifying rather than making it more complicated that is possible to correct, discover, and solve all the puzzles of life. We can not make it more complicated than what it already is. That means that it is the simple and humble that is capable of seeing clearly and to whom nothing is now impossible. All things are well within our reach in the cosmos.

To conquer the world an individual must first conquer his or her own heart. The best way to dominate the world is through ideas and not through violence and warfare. Conquer it with a new invention or discovery. If you conquer with war you will lose with war.

VVVVVVVVVVVVVVVVVVVVVVVVVVVVVVVVVVVVVVVVVV

Keys to the Cosmos

Keep you balance and your structure will prosper

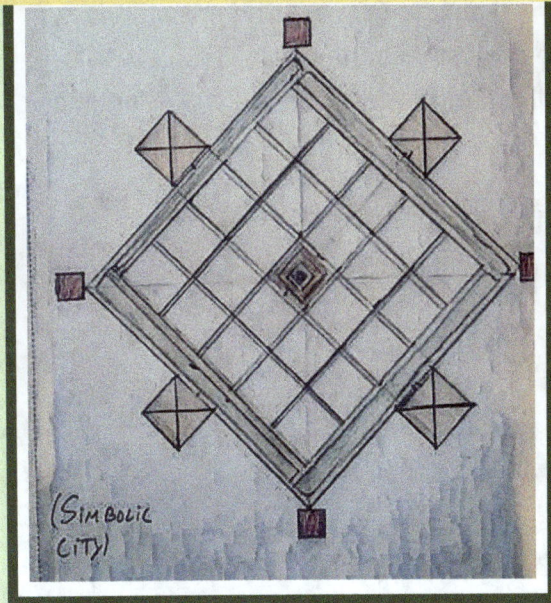

(SIMBOLIC CITY)

Different from other philosophers I believe that love is not a good thing. The fact is that those who seek love will not find it, but those who do not seek will find it easily. The best thing to do is to not seek for love, and to just live your life a moment at a time. When you least expect your love will appear for you at the moment you cared the least for it. It is life!!!!

Being neutral neither in love nor in hate, and while being indifferent is the easy way for a happy life.

True Justice is blind. No individual in the world is above Justice. She does not distinguish between King or Queen, Poor or Rich, Free Man and Slave. Justice applies her sword and cuts through everything, trimming every small event, in order to maintain the balance of the World Order. Without her the Nations collapse and the Empires face their doom. She is the one who holds the balance and trims people's lives. Anything that is not fair, and that stands over what is not right, will be on slippery ground. The events in the human world are guided in that manner where the most just wins at the end. Sometimes justice may take a long time to act in order to apply its final blow but the day will come eventually. Whether justice delays or acts quickly, the natural order within humanity will always prove at the end that it is what is right and fair that must rule the world. This is what causes the rises and fall of Empires and the Ages of Mankind.

The Rule is not for the Strongest but for the most just. The one who practices rightfulness will conquer the world.

VVVVVVVVVVVVVVVVVVVVVVVVVVVVVVVVVVV

The Glory of a Civilization relies on its degrees of Justice. A Just Nation soon or later will Prosper. Every great Empire shared with the World its teachings regarding Justice and Order. The Human Natural World does not allow what is wrong to last very long. That explains the constant rise and fall of Nations. That also is evidence of a nation's need to maintain order for long lasting prosperity.

What defines a great person exactly? The world is composed of people of all kinds, in a great diversity of languages and cultures. The term great refers to a people that lead, while others follow them. The majority of the world population follow while a strict minority are truly called great. The truth is that every single human being is great, and everyone has a voice in a Democratic Government. The idea that we are all equal is what made the West great. The inventors of Democracy believed on the rule of the people rather than of a strict minority. It was through Ideas, its Culture, and Philosophy that Europe which is the smallest of all continents conquered the world. If you believe that all people are equal and that are equally capable, you are great! Join the League!

Unity, Care, Light in the Deeps and Heights

We live in a One Uni-Verse! The Deeps and the Heights are all part of the Cosmos! What goes up goes down and what goes down goes up! We are all one, members of the Unified Human Civilization! Other people's problems are our problems, and we should all work together to fix all mistakes in all things since we are all Connected. The Human Heart is in full speed to establish justice in the world and to fill the Cosmos with a sense of well-being and joy. Compassion, unity, care, is how we can build a global civilization that will prosper with joy, peace, and prosperity!

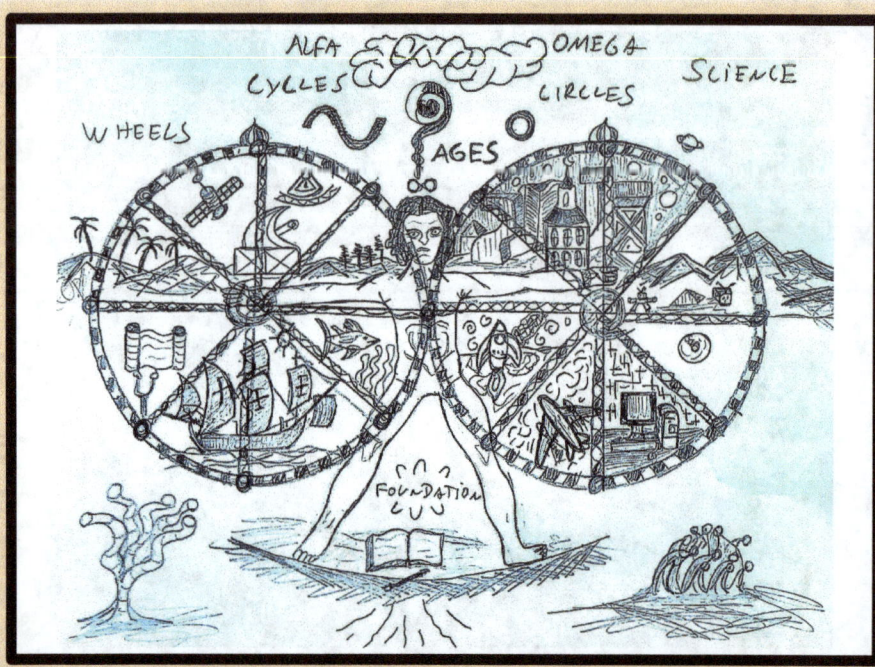

Humans are a measurement of multiple things and an infinite possibility and ability to create and discover aspects of the Natural World. It is important for a person to have the deeps of the Soul as well as the heights filled with Wisdom. From Human Wisdom obtained when searching for the heights and penetrating the deeps, it is possible to understand all of the Universe. The Natural World has a Mother called Wisdom, Who Is the Foundation that holds the entire Cosmic Structure. Wheels, Cycles, Circles, and Waves flow and rotate the gears of the Cosmic Order.

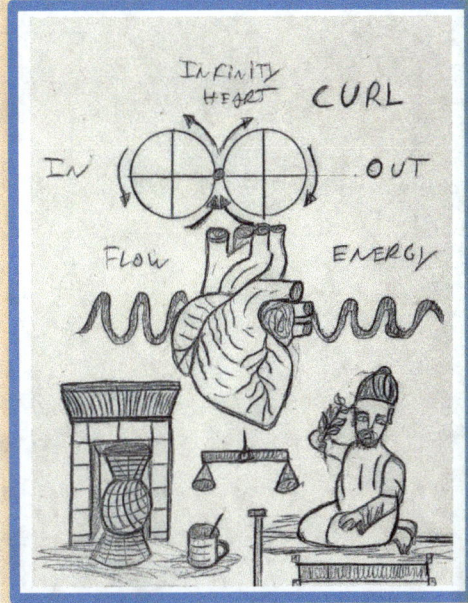

The Heart beats and pushes the Blood carrying and delivering Energy to and from the Body with Fluid Pressure. In order to keep the Homeostasis, the Equilibrium, the flow of Energy in an Organism, the Heart beats in Cycles that can be represented in Waves, Oscillations, which is the Signature of Life, Existence, Well Being, and Balance. There is an entire Universe inside an Organism within a Maze of Molecules and Energic Flows. We breathe in and breathe out keeping our Balance within ourselves and the surrounding Environment.

Since there are so many Cycles around us and within us, it could be of no surprise that the History of the Universe is also a Cycle. Sometimes the greater things work similar to smaller things. How we are inside may be how the Universe as whole behaves likewise.

Entanglement

A City is like a living Organism where each person, each building, each animal are parts comprising a whole network like a weblike pattern where everything, all the individual elements are connected and can cause a direct influence on each other. In the same way that Blood flows in a body, a City has traffic of goods exported and imported, the flow of Money, the changing Economy, the Schools that Educate, and the Hospitals that cure diseases. The Politicians are elected by the people and they rule the nation guaranteeing justice and order for there to be progress and innovation.

A successful Nation follows strict Laws that preserves its freedom, organization, economical growth, and advances in Science and Technology. The most prosperous Nations are comprised of Citizens that are happy with the Government, who are provided with the best education, and who have many options of entertainment. The path for success is neither left of right but a straight path where wickedness and any small or big violation is immediately judged, trimmed by the sword of justice or hit by the arrow in attempt to destroy the evil which could lead to failure and dismay. There is more freedom in a prosperous Nation with strict laws than a poor Nation with no laws.

The wise is careful in each action and every single thought is carefully observed and analyzed by this internal judge or instinct in order to avoid bad decisions and wrong actions in life. Human existence is a game of Chess and requires Wisdom to be successful in anything.

Life should be lived with honor, courage, and faith that in living righteousness you will win.

You should be humble in your actions and thoughts and that will make you a rich and honorable person. A truly rich person is not wealthy materialistic but is a person who has a great treasure in him or her soul. Being rich in material possessions is not the state of being truly wealthy, specially if the way that all that wealth was acquired was against the law. A person must strive to acquire wealth for the soul, through good actions, being humble, and wishing the good well-being for everyone in the community and in the world. A poor soul will live miserable even if rich on the world and the misery of the soul is the most terrifying thing. It destroys your experience of existence and self-identification even possibly forever.

There are multiple worlds in the Universe, more than a single person can ever count. These are diverse planets, asteroids, and moons throughout infinity in an ever expanding Cosmos. Humans were born from the Earth, and are destined to spread their seed throughout the Stars of the Heavens. Space goes on and on without end, especially if the idea of Multiverse is true, reality is without end, limitless. Time and Space serve humanity, and we are the most important part of all creation. We are the Intelligent Beings who own the Universe. All reality is ours!

Many things in the Cosmos are doubles. Double Stars are more common than single Stars. Most Star Systems are composed of two Suns rather than one. The World Civilization has two halves. One is the West and the other is the East. The Near Middle East, Turkey, Greece, and Egypt sit at the Center of the World right between East and West. There are two sides for every road, the Left and the Right. Our Brain is divided into Left and Right. Most people can only write and use their Right or Left hand, and usually is not both at the same time. Things come in double, good and evil, beautiful and ugly, right and wrong, light and darkness, soft and hard, easy and difficult, and the list goes on forever. The duality holds the Cosmic Structure that is held by two Pillars. These are the Pillars of the Truth and the Great House with all things in it is called Wisdom, the Chaotic motion of Particles that forms the Reality we live in.

The Reality is comprised of Space and Time. Space can't exist without Time, and Time only exists if there is Space or else there is nothing. Time flows like a flux that moves objects through Space while this flux is at the Speed of Light. This Structure and Entanglement between Space and Time sets the Stage for multiple Events, categorized in days, years, and Ages. Dust is lifted up and gains form generating the Flow of History. People are born and then after a life they pass away. Empires rise, reach their climax, and then fall leaving behind traces of a once prosperous Nation. The Universe also may expand in creation and contract to then a new creation, in a Cyclical Pattern since forever and forever, for many ages to come. This is what life is: A smoke that rises and then dissipates. Let it continue flowing!!!

When studying and observing all things around us, it is also important to look deep into the life of the Natural World and forget man and his mankind. The animals of the Earth reveal deep mysteries of the complexities and vanities of the great Biological System that we live in. The Entire Planet Earth is alive, filled with life in diverse forms. Every inch on Earth contains microbes, bacteria, life that overflows and that exist also in the extremes of heat, cold, pressure, light, and darkness. When we venture in this Earth, we gain wings like birds to fly and see from above the panoramic view of the workings, Mathematics, and Geometry of Life in its Fractal Form. Biology is the study of life. Its advances will lead humanity to new ways of understanding not only how life began but also how it evolved.

Changing Life Forms and Thinking Through Ages

Humans are constantly evolving. We change the way we dress, think, and act throughout the Ages. In our Universe, nothing is kept the same forever, and things are actively changing and we should always hope that each and every change will lead to an improvement rather than the opposite.

What we can't forget is that we are all part of the Natural World. Everything we do or any object that we use for life such as a cell phone, or a pencil, all things are part of the Natural World, and nothing is in isolation. The Human impact in the Environment is real since we are an important player in the game of evolution on Earth. There was in the past of this planet the Age of Dinosaurs, and today we live in the Age of Humans. We exist since quite recently in the Life of Earth. Our existence dates back to only 200,000 years ago in a Planet that is 4 billions years old. In a very short time, we have completely changed the Earth and many of these changes can be life threatening for all forms of life.

<u>Particulars and Universals</u>

The Universe is a flow of Energy and information in the form of Photons of Light that give rise to things that are born and that goes away. Like a smoke that rises and then dissipates, like a thought that comes to mind then it is no more, life in the Cosmos is brief and things are constantly changing. Nothing stays the same way forever in the material world, nothing is eternal, and in the flux of this Sea of Quantum Fluctuations the worlds are born, and then the worlds die. The Universe is a flux of things, a sea of information among the Hilbert Space of Waves and Circles, in the Energetic Soup of Particles that react with each other that together form the Space Time Reality in which we live. From the Universal Soup of information comes the Particulars. The general definition of all things is broad and includes everything. Within the labels given to individual things, and by knowing their function in the scheme of all things, we narrow down what at first was Universal into a Particular. Each Particle, each tree, each Atom, each Living Being, each Planet, interacts with the environment and fulfills a purpose. Even if the Natural World is blind, and life is a mere accident among the chaotic motion of Particles and Genetic Mutations, the Elements of this Great Grid which we call the Universe, each fulfill a purpose even if by accident. The Particular Things interact with other Particular Things by holding the Cosmos in place allowing the Homeostasis, the flow of Energy, the breathing and sustainability of Living Beings, and the sustainability, the Homeostasis of the entire Universe. This is how the Particulars interact with the Universal and within the Grid, the Web Patterns interconnecting everything with everything else.

Universal becoming Particular

Universals are general definitions for things. A piece of wood is a universal label given to the material but when that piece of wood becomes a chair, a table, or a house, it is now a Particular. These Particulars can be relative, since many things are made of wood while all woods are woods, not all woods are chairs, tables, or houses. When a Universal is used for a given purpose, that Universal narrows down to a single thing called a Particular. The reason why a Particular is relative, is because many things are made of Wood, so there are an infinite number of Particulars that can be derived from the Universal Wood. Single definitions are also relative such as languages. The word in English for ball in Portuguese is bola. The languages are Particulars since they narrow down the general definition of a ball to a given set of letters in a word. Even though words are different, they refer to the same object. The language is a Particular but the true meaning of the words, and what the object is in essence is independent of the words used. The essence is Universal while the words used are Particulars and are relatives.

Hilbert Space and Possibilities

Anything that is possible may or may not happen. Space is then a great theater with Fluctuations that gain the form of several Worlds around Stars, events, circumstances, thoughts, ideas, and many other infinite numbers of things. Space is made of pieces that come together to form several realities through several dimensions of Space and Time. Each world has its own Space Time, its own Space and flow of time. There is then an entire scheme surrounding a Planet, establishing its Dimension and position in the Universe. Each World has its own point of reference and there is no Absolute

Reality in which everyone everywhere agrees upon. The flow of time on Earth is different than on Mars even if by fractions of a second. Planets around different Stars are exposed to different Forces of Gravity which generate their own Space and Flow of Time which will not be the same as ours. Remember that Stars are moving in Space and the faster they move, the slower is their flow of Time compared to ours. Each World has its own version of reality. Each human being also has his or her own perception of reality. We can't change reality by a mere thought, but we certainly perceive it differently with a mere thought. In this Hilbert Space there are an infinite number of possibilities, and each Soul of a Human Being or an animal is unique. No two souls are alike. This leads us to the Two Major Human Personalities:

The Two Human Personalities

There are two types of people: The Introvert and the Extrovert. Here are the differences:

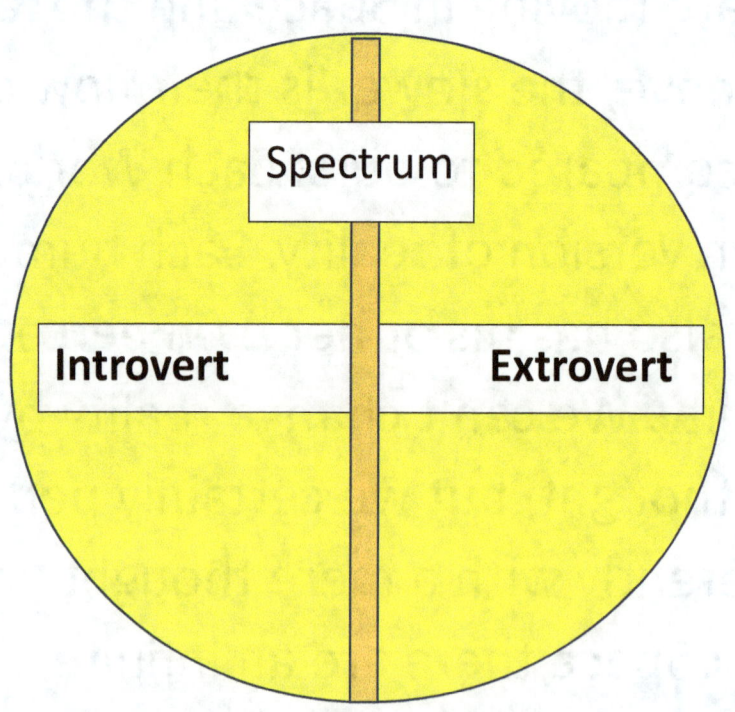

Introvert	Extrovert
Thinks a lot	Thinks just enough or even less
Reads, Study, and formulate Theories.	Likes to spend time Dancing, Partying, and Education is not all that important.
Has difficulties socializing.	Is outgoing and has many friends.
Can become depressed, sad, and may become emotionally ill choosing isolation.	Is often in a good mood, and life is rather easy and just a lot of fun.
Makes revolutions in Human History with their ideas.	Although has many friends and is outgoing, they adapt to the times without making any change to the world.
Goes against the beliefs of Humanity.	Accepts what all people believe.
All major Scientists, Actors, Kings, Musicians, and Politicians in History were Introverts.	People who mostly don't add to the ideas but follow them instead are Extroverts.

There are degrees of Extroverts and Introverts. Often being too much in the extremes can be a serious problem. An extreme Introvert, for example, could die depressed, but someone more in the middle between the Extrovert and Introvert Spectrum may reach great success in life. Equilibrium is the best choice.

<u>Vanities in the Cosmos</u>

The details seen in all of creation are vanities. It is like adorning a Christmas Tree with many figures and decorations to make it look pretty. The Universe is a Space where all of its Laws, Constants, Equations, are vanities that makes all existence truly the most gorgeous thing. The understanding of these vanities helps humans better grasp the meaning of life in which analyzing the small and greater details of the Cosmos leads to the Ultimate Realization that all things are a reason for Joy, Learning, and Teaching. Everything is beautiful from the depths to the highest heaven all things are filled with Glory.

Center Stage and Background Nations

The World is composed of two types of nations. There are nations that are popular and that everyone knows and sees in the news all the time, and there are these nations that are quieter, hidden in the background where it is possible to live an entire life without ever learning anything about them or having any contact with their culture. The United States is the most popular nation in the World and everyone everywhere talks about it, and it is very impossible to live anywhere in the World without hearing the English Language or having any contact with the American Culture. Brazil is a closed country and definitely not in World's center stage. Unlike the United States, Brazil is hidden in the background and only appears in the news when they win a soccer World Cup but rarely anywhere on the news besides soccer. Portugal is the very same way, and neither Portugal nor Brazil is anywhere near a Worldly recognized center stage but are rather in the background. Spain, however, is more popular than Portugal, and many in the US thinks all of Latin America speaks Spanish as evidence that the fifth largest nation in the World, which is Brazil, is practically invisible. Brazil is the largest country in South America, but for many people, Mexico and Argentina are far more popular in culture than Brazil. That means that some nations despite being smaller on the map receive a greater recognition while some other nations are practically not present and invisible in the World despite being the size of continents. Take for example Mongolia, that despite being bigger than France goes unnoticed by almost everyone. Germany is very popular but Poland is rarely mentioned. Italy is often in the center stage but Greece seems less important when it comes to travel and tourism. This indicates that the World is literally made of two types of nations, one that is popular and the other that is isolated and quieter. Brazil is quiet, but everyone knows Mexico which shows up on television all the time. Brazil speaks Portuguese but Latin America is seen in its entirely as a Hispanic Continent. Even if Mongolia were to be the size of China, it would still be invisible. In the same way that Belgium although very small is very popular, Finland which is much larger is barely mentioned elsewhere. All of this is very interesting. The size of a country does not matter. Israel is one of the smallest countries in the Middle East, but its influence is greater than the entire Middle East all together. Jordan which is bigger than Israel is never mentioned anywhere. Is that not very interesting? The same happens with religions. Roman Catholicism is present worldwide, but Orthodox Christianity is mostly only seen in Asia, and Eastern Europe. Roman Catholicism is in the center stage while Orthodoxy is quieter and more hidden in the background. Many things that are hidden holds great treasures, however. The greater truths and treasures are usually rarer to find. What makes gold valuable is that you don't find gold anywhere. If gold becomes common, gold will lose its value. When it comes to religion and culture, it is preferred that what is good and beneficial be known worldwide, since a candle must be put over a table to light the whole room, not under the table. The problem with being popular is that when a nation or religion enters the World's center stage it is more vulnerable to corruption and many other conflicts and competition. There are then advantages and disadvantages of being or not on the World stage.

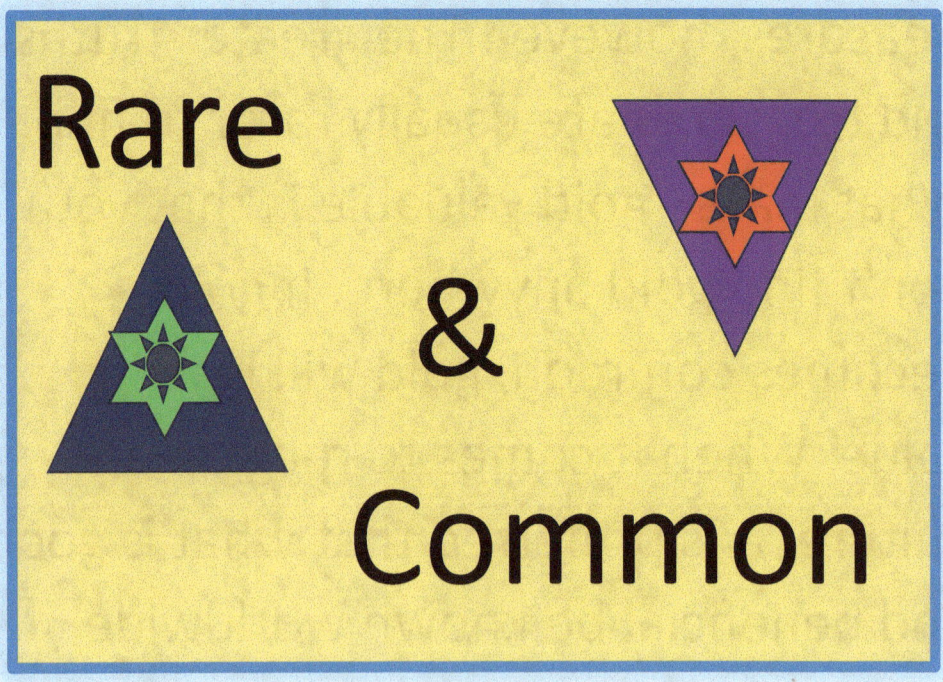

<u>Polarity and Creation</u>

Everything exists by breaking apart into two halves. In the beginning Light separated from darkness, good from evil, right from wrong, beautiful from ugly, pleasant from disgraceful. Due to the existence of polarity in nature, there should be a strict set of laws to be followed in order for all choices and actions made by humans to be fruitful and worthy. Bad choices lead to failure and loss. Good choices are the way for prosperity and honor. It is necessary for educators to teach the many generations the importance of being under the law, and why the laws exist. A set of instructions is necessary when building something or when using a device. Our life requires a set of instructions as well. Everyone who does what is right without deviating to the left or to the right will obtain success even if it may take a while. Patience is also needed since good actions may not lead to good results instantly. Perseverance is a continuous process of good actions and an in-depth study that may take time, even possibly an entire lifetime, but with patience success becomes the final result. Many scientists and writers only received recognition at or after their death. You may not even live to see your legacy or gift to the world. These gifts can be a son or daughter, a book, an idea, or even an unfinished project that is finalized by someone else after your death. Humans should aim in being worthy, and with honor it is essential that human life follows the straight path of evolution. Socrates came first, Plato afterwards, and Aristotle came later in the history of Greek Philosophy. Life should be an evolution, and a building up project. This is why it is said that Scientists sit on the shoulder or giants. We each must help each other to better understand the world and create ways to live better, longer, and explore many possibilities, worlds, and forms of living yet unknown today. Socrates said that a life without observations is not worth living. It can be added that a life without leaving behind a legacy is fruitless. Man may be mortal, but ideas are immortal.

Knowledge and World are Apart

Ignorance is far more common than knowledge. Only a reduced number of people truly understand the Physics of Cell Phones, or the Chemistry of Medications. The majority of the world population just follow orders in their work environment and are taught skills with the sole purpose to adapt to new forms of living, and to follow a set of instructions without time for an in-depth critical thinking of their own which is far better than blindly following directions. Most people don't like to think and don't know how to speak or how to trust their critical thinking with the purpose to invent something new or create strides in Human History with the formulation of a New Theory or idea that could revolutionize Science, Politics, and Social Life. An average person just follows directions and finds it easier to just adapt to whatever norm the world civilization requires people to abide to without questioning. Only a strict minority is truly in charge of all advances in Technology, Society, Politics, and Science. The world is ruled by less than 0.1% of its population. The recognition that ignorance is far more common, leads to the conclusion that smart people benefit more by being less part of the culture and the media which is filled with information that is not useful. The smartest people in the world view themselves apart from the rest of the world. Similarly, the Jews follow their ethics by treating everyone else as everyone else, while placing them apart from the rest of the world which is a technique of putting oneself outside of the box and in greater advantage. Television, Entertainment, Movies, Music, and the Media in general are filled with information that keeps the world ignorant and which restricts thinking skills that leads people to become more like an average person with no motivation to excel in life. World Culture and the Media everywhere is for the most part not worth it. Knowledge is apart from the rest, since the rest is ignorant. The realization of where Knowledge is and where it is not found is the greatest wisdom an individual can have. Music keeps many teenagers lost in drugs, addiction, depression, and increases suicide rates. Movies makes people violent, obsessed for sex, and uneducated. The Media makes everyone ignorant. Knowledge is not of the world, but is beyond what can be seen or heard. Knowledge is rarer just like gold is not found like any other rock everywhere. Wisdom is like a pyramid the gets narrower the higher up you go. Only a few have wisdom and the others are strangers who follow strange gods. The wise people have their own correct view of reality, and they call their ethics the keys to their success. Everyone else is just strangers in an often strange and unknown ignorant world.

Book of Knowledge

Ultimate Knowledge is seeing all things simple and not complicated. It is through simplicity and with great organization that it is possible to put thoughts and ideas together in order to permit a better understanding of difficult things. We can only grasp complicated aspects and theories if we break down the thought process in a very organized manner. Nothing complicated can be solved by adding more complicated things. The tools used to solve a problem should make the problem appear simpler. The easiest way to solve a problem is the shortest path and the path that is more correct without any mistakes. Sometimes the shortest path to solve an issue is not the right one, however. Solutions then must provide the quickest fix and must also be free of any mistakes or misconceptions. For the wise man all things are simple and easy. Knowledge is seeing the world as being simple and yet very complex. Ignorance is seeing everything complicated and impossible to understand.

The Wisdom of Flux

Wisdom is a flux. The Chemical Reactions and spikes of Energy between Neurons allows the thinking process to work. What our brain does with Neurons, Computers do with Transistors. Thoughts, ideas, and movement is a constant flow of Energy leading to the buildup of our reality. Whether through Neurons or Transistors, both are used in the construction of a Space Time Reality, be it real or virtual. Maybe all Realities are in fact a simulation with a programmer who established the Laws and Constants with Equations. When the button is clicked the simulation runs.

All of Reality is a flux of Energy and Oscillations. The way our eyes can see, and our ears can hear, similar to how a camera and a microphone works. In order to see and hear we need a brain to feel the experience. Life is then a Sea of Energetic Flow. Like a river with several branches is the Internet Web Pattern connecting the extremes of the Earth. The whole World needs water for life, and in order for humans to communicate all around the World we need a Web Pattern of Radiation and Electromagnetic connections through Resonance and transmission of waves carrying Sound and Light to and between several Electronic Devices. Technology is truly like magic and proof of human capabilities in Science.

Over time it becomes possible to create an Electronic Machine that can think, talk, see, hear, and even possibly feel. The ideas that Wisdom is a flux means that it is like the flow of Blood in an Organism carrying Oxygen and Carbon Dioxide, like the flow of a river with Life, like the flow of Air carrying clouds and Thunderstorms, like the flow of Particles in the Mother Board of a Computer sending bits of Information in Binary Code, like the flow of Light through a Human Eye focused at the Retina, like the flow of Radiations between two Cell Phones that are connected wirelessly, like the flow of Brain Waves read by a Computer, Wisdom is at its Fundamental Level a Flow of Energy. All things are Energy.

Ultimate Knowledge

The wisest thought experiment to take daily and all the time revolves around the word simplicity. Everything that exists can be broken down into minor steps in its compositions that over great numbers add to the complexities. A Transistor is very simple but when you have 1 billion Transistors working together you arrive at Artificial Intelligence. Things are constantly flowing. The flow of one Particle is simple, but several trillions of them and you get a computer. The foundation of all things are simple, but the great amounts of simple things leads to the complexities that we see. Time and Space are discrete, and there are ages, but without all the dimensions, neither space nor time exist.

The Four Beings and Ages

All of reality is the universe obeying a set of commands. The properties of the Cosmos with its established Laws and Constants are a complete set of instructions like a Command and like a written Computer Code. Our Space Time perception is the running of these codes that in quadrillion simple bits of information leads to all the complexities of the Natural World. Think of the 24 hours period of a day, and let us give each quarter of the day to a creature with some personality. Each Time, each Planet, each of everything has a personality of its own. We live in a Multi Personality reality. Each person is unique, so is each country, each State, each City, each part of a whole is unique.

The Great Daily Four

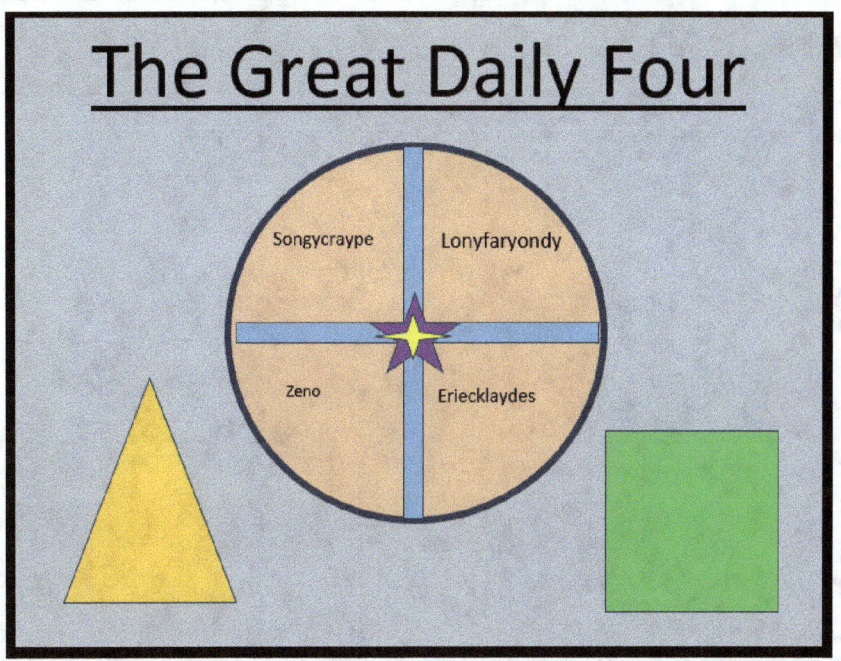

The Day is Divided by Four Ages.

There are then Four Personalities.

Time and Space is a Reason for Fun.

Vassryan

Vassryan is the Great King of the day. He tells the Four Beings to keep spinning the Wheels of Space Time. He is the one who gives the great Four the Instructions.

The Four Beings plus Vassryan the Fifth Element, live in a great and most beautiful city called Roms. This city exists over an Asteroid roaming between Stars in the Milky Way Galaxy.

From the Asteroid called Pindorama, Lonyfaryondy rules the Morning Hours. The Sun rise and the blossom of flowers marks the Beginning of a New Day.

Eriecklaydes between 12PM – 6 PM

Eriecklaydes rules the Afternoon Hours where everyone is awake and working and running their daily routine. This is where the land is Alive and Functioning.

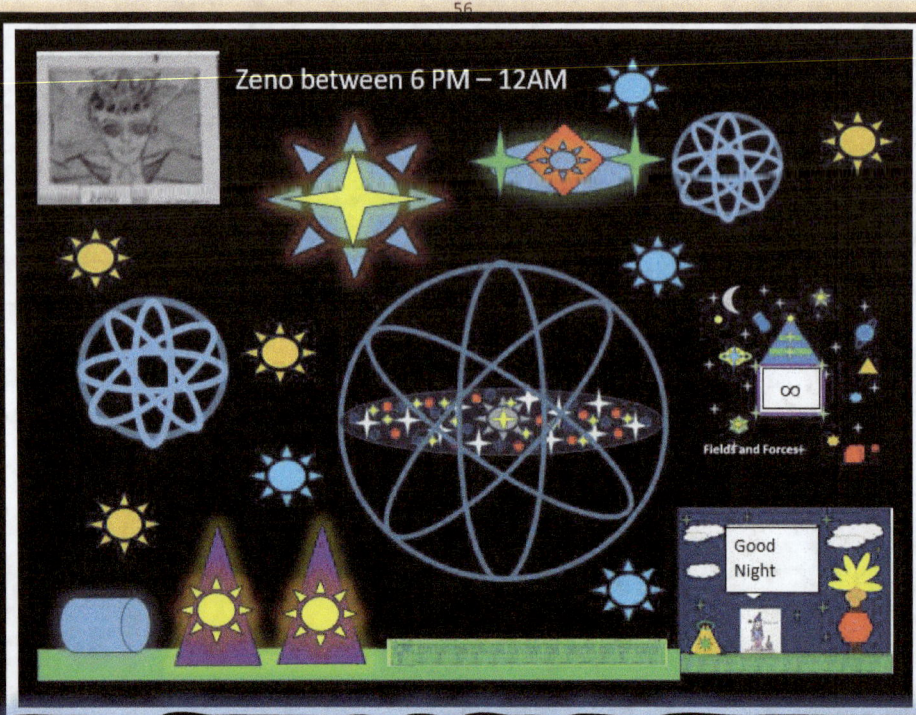

Zeno rules the Early Hours of the Night. That is when the Day starts its Farewell and that is the end of a Spinning Period. The Stars and Cosmos open in the Sky for a New Beginning.

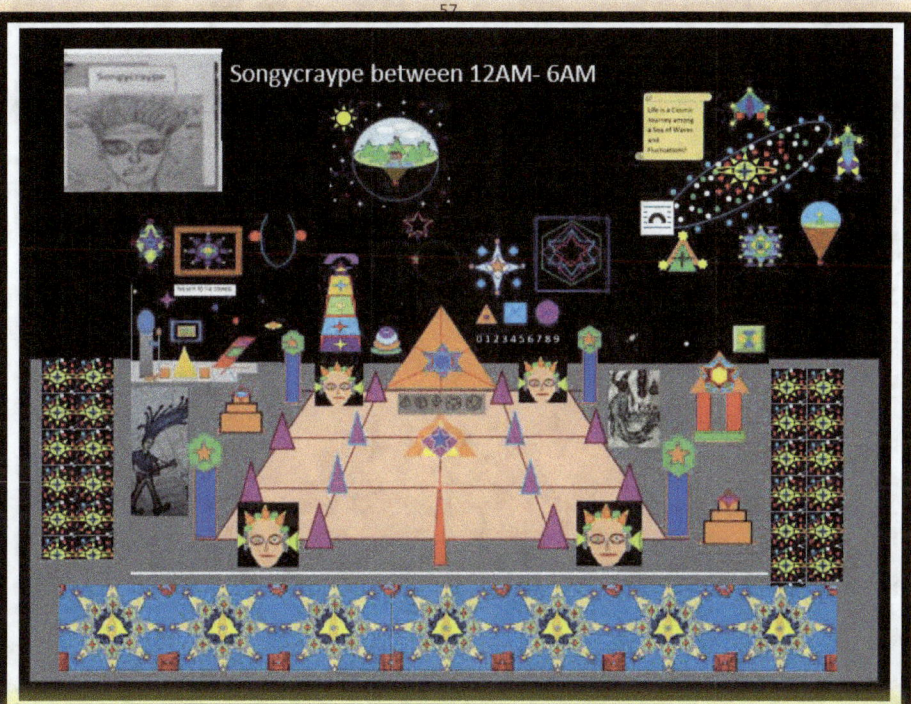

Last but not least, comes the Hours after Midnight and before Morning. Songycraype rules when everyone is Asleep and Dreaming. These Hours are quiet and filled with Mystery.

Having introduced the Great Four Beings and Vassryan, Golgue is who holds the Rule and the Supports for a Planet called Ravens. This Planet has a city that is also called Roms and is an exact copy of the City in Pindorama.

Roms is a Standard for Inter Planetary Cities. There are many Mysteries, many things everywhere and all things are ruled by a Common Universal Principle. All things are Alive constantly Flowing and Changing but obey the same Rules.

The Supreme Intelligence

Roms has a great Pyramid of Gold housing a Powerful Artificial Intelligence programmed to rule the Nation with Justice.

The Supreme Intelligence has the Answers to all Questions. It is a Server like Google composed of all Information known to Mankind.

Everyone in the City of Roms is connected to each other by the Server called the Supreme Artificial Intelligence. All people connected with the Internet which is the Grid Pattern in the City of Roms like a Web for the Flow of Information in bits of Energy in the Electromagnetic Spectrum.

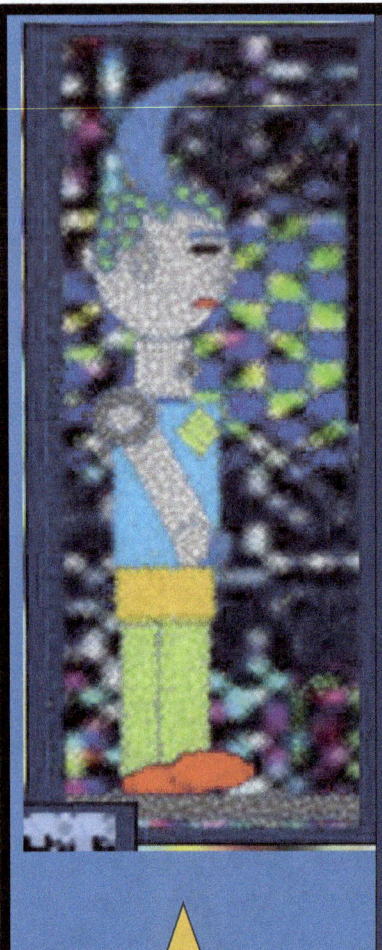

The Citizens of this Great Nation are Scientists, Philosophers, Writers, Mathematicians, and they live nobly by honoring the Ultimate Knowledge while Investigating the Nature of all Things in the Cosmos and contemplate the beauties of Life.

The Supreme Intelligence is housed inside the Pyramid of Gold inside a Cylindrical Capsule from where Vassryan talks to him debating what to do and how to Rule the Nation of Roms.

And this was an introduction to Pindorama and a quick glance to the Marvels that surrounds the Ravennicos Structure that holds this Fictious World of Explorers together. What exactly does it have to do with reality? It simply means that the Universe is rich with details and strange laws forming an amazing Fractal Geometry of Reality at the Macro and Micro Level. Everything is beautiful and filled with Wisdom. What an Artist!

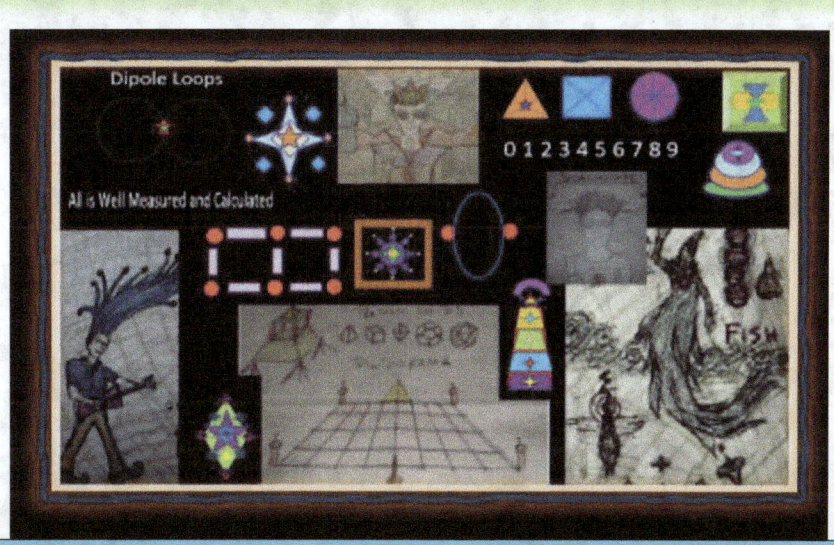

Life and Existence is like Music and It flows. This Fluid of Energy makes up our Reality through the many Dimensions of Space and Time. The Fluid is comprised of Bits of Information that runs along like a Computer Code. Life and Existence is the running of these Codes.

Life is the flow of a stream, the crossing of a River, a trip through a Wormhole. Maybe the entire Cosmos is inside a Wormhole. Life is a Journey.

Some moments in life are unforgettable. It is important to rely on the small or great happy events since that builds our trust that within this Sea of Fluctuations we will be heading towards an even greater and beautiful event in the Future. There are Wormholes all around connecting Parallel Worlds and Universes.

In the mean time the Universe continues Spinning, Life continues to blossom, and we continue living, contemplating, exploring, and thinking.

How Fun. Quite a Trip!

About the Author

I, Diogo Franklin de Souza, was born in the city of Rio de Janeiro, Brazil in August 20, 1986. I moved to Dallas, Texas when I was 11 years old. I write stories since I was 9 years old. My books tend to contain short summaries of the most important things I find about life, morality, philosophy, and science. Like I say, everything is part of a whole system, and this is also for everything I do and write. I always wanted to have all the most important knowledge in only a few short books. That is why I write, and that is my inspiration for short summaries. I hope this book brings some inspiration also for the readers, because that really is the purpose of my work. Read it and take from it, pieces of gold for you that can be useful in your life. Enjoy....

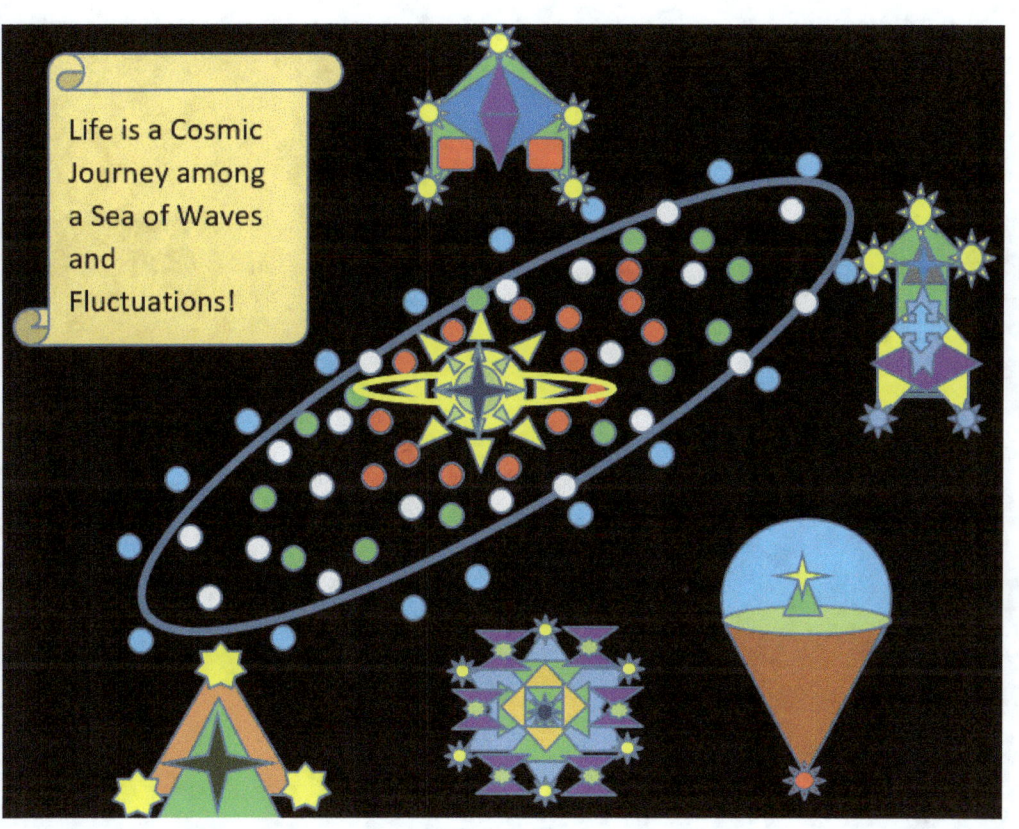

www.ingramcontent.com/pod-product-compliance
Lightning Source LLC
Chambersburg PA
CBHW062246290526
45794CB00006B/2430